Centerville
Washington-Centerville Public Library
Centerville, Ohio

DISCARD

W9-BMI-610

Inventing the Christmas Tree

Inventing the Christmas Tree

Bernd Brunner

Translated from the German by Benjamin A. Smith

Yale UNIVERSITY PRESS

new haven and london

An earlier version of this book was published as *Die Erfindung des Weih-nachtsbaums,* © 2011 by Insel Verlag, Berlin.

Copyright © 2012 by Bernd Brunner.
All rights reserved.
This book may not be reproduced, in whole or in part, including illustra-tions, in any form (beyond that copying permitted by Sections 107 and 108 of the U.S. Copyright Law and except by reviewers for the public press), without written permission from the publishers.

Yale University Press books may be purchased in quantity for educa-tional, business, or promotional use. For information, please e-mail sales.press@yale.edu (U.S. office) or sales@yaleup.co.uk (U.K. office).

Designed by Sonia Shannon.
Set in Stempel Schneidler type by Keystone Typesetting, Inc.
Printed in China.

Library of Congress Cataloging-in-Publication Data
Brunner, Bernd, 1964– [Erfindung des Weihnachtsbaums. English]
Inventing the Christmas tree / Bernd Brunner; translated from the Ger-man by Benjamin A. Smith.
    p.   cm.
"An earlier version . . . was published as Die Erfindung des Weihnachtsbaums, c. 2011 by Insel Verlag, Berlin"—T.p. verso.
Includes bibliographical references.
ISBN 978-0-300-18652-9 (cloth : alk. paper) 1. Christmas trees—History.
2. Christmas trees—Germany—History. I. Title.
GT4989.B78 2012
394.2663—dc23   2012012737

A catalogue record for this book is available from the British Library.

This paper meets the requirements of ANSI/NISO Z39.48–1992 (Permanence of Paper).

10 9 8 7 6 5 4 3 2 1

 Contents

Inventing the Christmas Tree

A Tree Full of Mystery

A colorfully decorated, fragrant tree, lit with strings of twinkling lights—the most important and enduring symbol of Christmas, the major annual festival in the Christian world. When it's snowy and dark outside and the days are short, the tree lets us dream of nature's life force returning. Its deep evergreen is the symbol of life long-lasting, and its decorations—anticipating the buds and blossoms of the coming season—give the tree a hint of fairyland. A visual magnet, it lightens the gloom of winter, delivers a slice of the forest within the walls of the home, and, on Christmas Eve, awakens joy and hope that the sun will soon return for longer hours.

Sometimes it can be useful to make the familiar seem alien. Of course, trees are not normally found in houses, nor are they decorated with candles, straw stars, tinsel, or glass balls. Instead, they are more likely to bear blossoms, fruits, or pinecones, or the weight of birds and squirrels. And most trees have roots that stretch deep into the ground. But the tree in this book usually has only its crown and trunk—and its place is in the home. A wholly remarkable tree, in short, one in need of an explanation. Although now inseparable from Christmas for us, the tree isn't found in this form in early Christianity at all. It is missing even from

Isn't this tree a bit too big?

church songs and prayers. Something mysterious surrounds this tree that first appeared in the German cultural context, before it literally branched out into other central and northern European countries, America, Russia, and beyond, capturing the imagination of many people. This book also tells a small global story.

What drove people to go off into the forest, chop down a tree, put it in their house, and decorate it in the first place? Is it really just a pagan remnant—as conventional wisdom has it—or is the history behind it more complex? What is the symbolic message it conveys?

The Search for the First Tree

And then, all of a sudden, there it was. It seems to have appeared *ex nihilo;* first only here and there, and soon all the more frequently. Scant notes, censures, and prohibitions in yellowed documents and notebooks of old testify to its existence, but even these are mostly indirect. A precise record of its appearance is nowhere to be found. In 1419 the Freiburg Fraternity of Baker's Apprentices appears to have seen a tree decorated with apples, wafers, gingerbread, and tinsel in the local Hospital of the Holy Spirit. Another document claims that the first Christmas tree was erected in Tallinn, Estonia, in the year 1441. There the tree was set up in front of the town hall for a dance. The record is ambiguous, though, for the Middle Low German word that was used—*bom*—could also have referred to a decorated mast or pole.

In Riga, Latvia, the claim is that the first decorated Christmas tree can be dated to 1510. The so-called Blackheads—foreign traders who had formed a guild—are said to have erected a tree in front of the town hall at the time of the winter solstice. Children decorated the tree with woolen thread, straw, and apples before burning it at Lent.

Even in England, during the reign of King Henry VIII (1491–1547), a special kind of seasonal tree seems to have been prominent, according to the Loseley manuscripts:

> Agaynste the xii daye, or the day of the Epiphanie, at nighte before the banket at Richmonde, was a pageaunt devised like a mountayne glisteringe by night, as tho' it had bene all of golde and set with stones; on the top of which mountayne was *a tree of golde, the braunches and bowes frysed with golde, spredynge on every side over the mountayne with roses and pomegarnettes.* (quoted in *The Living Age*, August 8, 1857).

Is this an early Christmas tree?

These are not the only conjectures. Muazzez Ilmiye Çığ, a Turkish archaeologist, believes that the Christmas tree has its roots in the Central Asian steppe, where people covered a wishing tree in ribbons in homage to the god Bai-Ulgan, who lived above the sun, the moon, and the stars. This offering was made in the course of a festival celebrated by the ancient Turkish folk on December 23. The custom then apparently was brought to Europe by the

Huns. This theory is disputed and few agree with Çığ that this tree ritual gave birth to the Christmas tree we decorate today.

Let us return to central Europe, for only there is the continuity of the tradition's first few centuries undisputed. Above all, we must look to Alsace—that rich Franco-German landscape between the green slopes of the Vosges Mountains in the west and the Black Forest in the east. There a Christmas tree, raised in the Strasbourg Cathedral, can be dated to the year 1539, a time of economic blossoming, before the Thirty Years' War.

The chronicle of a guild in Bremen from 1570 contains references to a tree placed in the guild's hall and decorated with apples, nuts, pretzels, and paper flowers. For the Christmas celebration the children were allowed to shake the tree as they would have during the fall harvest. Sometimes these decorated trees were apparently carried in processions and the poor were allowed to plunder the fruits and baked goods before everyone began to dance. In essence this was a continuation of the pre-Christian "fruit trees," as they were called, which bore no candles.

Where the first tree stood is lost to the ages. But we can assume that these more or less random extant documents refer to something that was already in existence decades before. What is certain is the appearance of the trees in the trade guilds of the sixteenth century.

But even a bit earlier, tree felling must have become so

BAÜMLI, BAÜMLI, SANICKLAUS-
BAÜMLI.
Die Eltern find der Sanct Niclaus;
Ein Efel heißt der Dienst im Hauß.

Morality Tale, by the Swiss engraver David Herrliberger (1697–1777)

prevalent that in 1494 the Strasbourg jurist and town clerk Sebastian Brant found it necessary to forbid the custom of cutting off pine branches at the New Year and bringing them home. In 1554 felling trees for Christmas was officially banned in Freiburg, in the region of Breisgau. In Upper Alsace, where trees were apparently more plentiful, a more lenient method of limiting the custom was attempted in 1561. Every citizen could take from the forest no more than "one pine in the length of eight shoes." It is clear that the custom was known and loved in Alsace and the surrounding regions five hundred years ago and more.

At this time—toward the end of the sixteenth century—appears an early form of a Christmas song that has since become a beloved standard. "O Tannenbaum," written by the evangelical composer Melchior Franck, was not used in the context of Christmas until 1824, when it was sung with lyrics by the Leipzig teacher Ernst Anschütz.

An image by the Weimar court copper engraver Carl Schwerdtgeburth, which he created for a children's book in the middle of the nineteenth century, shows Martin Luther with his wife, Katharina, his children, and his parents sitting beside a Christmas tree. Was this a cozy family gathering that actually took place as shown here? There is also an account of Luther walking alone in the forest one

An artist's imagining of Martin Luther and family with their Christmas tree

night and being inspired by the stars to place candles on a small evergreen tree. The church reformer lived until 1546, and it is known that he encouraged the celebration of Christmas, but the first confirmed Christmas tree in Wittenberg—Luther's hometown—did not appear until the eighteenth century. Family celebrations involving the Christmas tree are also a much more recent phenomenon. We now know that such celebrations did not take place until the end of the eighteenth century, in part because only then did bourgeois family life create the conditions for such an event. But history has limited power over popular perception. In fact, Schwerdtgeburth's image circulated so widely and was reprinted in so many publications that it took on a life of its own, and actually helped make the tree more popular among Lutherans.

Mythical Trees of Years Gone By

The search for the first Christmas tree is a quixotic quest. Trees have always been part of human life, both practically and symbolically. Rock art found near Bohuslän in Sweden depicts individual evergreen trees in addition to other motifs. Whether these drawings point to Bronze Age tree worship is uncertain. We do know, though, of the holy sun tree mentioned in ancient Indian texts: it shoots out of the ground at sunrise, grows so tall during the day that its branches touch the sun, then slowly grows smaller until, at sunset, it again disappears into the earth.

In the Edda—the ancient Norse epic—the world tree Yggdrasil symbolizes creation. In Uppsala, Sweden, according to old manuscripts, a holy evergreen yew tree stood near a temple dedicated to the Nordic gods. A spring is said to have had its source at the same place. Sun trees, world trees, and trees of life also thrive in legends of many Indo-Germanic peoples, remaining as vague traces in the collective memory. Greek mythology, too, is rich in examples of deified trees and humans transforming into trees.

The world tree in an Icelandic manuscript from the seventeenth century

9

Trees, the largest members of the plant world, pervade human thought.

Think too of the beautiful fruits on the trees of Elysium, the garden of the gods, as well as the golden apples of the divine nymphs, the Hesperides. These apples were considered the finest jewelry and the most delicious of fruits. They were the food of the gods and were said to bestow immortality—which is why mortals were forbidden from eating them. A similar proscription is familiar from the book of Genesis, where Adam and Eve are forbidden to eat from the tree of knowledge of good and evil, "for," God warns them, "in the day that you eat from it you will surely die." They eat, of course, and are exiled from Eden and burdened with mortality and all the other afflictions of the human condition. So prominent in the creation story is the forbidden fruit that it is easy to forget the other tree nearby in the garden, the tree of life. In Christian symbolism, this tree has been associated with the cross upon which Jesus died. It is seen as a symbol for communion with God, and in some later variants of the Bible stories, its fruits give immortality. According to some accounts, the forbidden tree and the tree of life are one and the same.

Outside the realm of myth—but nearly as fantastic a bearer of life—is the slender, tall palm tree, which can live to a hundred years and often bears a rich bounty of dates. It is well known in both the Occident and the Orient. The tree of the Holy Land and of Arabia is found persistently in

Christian depictions of the stall in Bethlehem, implying great significance, for time has pruned all inessential elements from these stories. When the Bible was translated into modern European languages, many motifs from the environment of the biblical stories were adapted to the local conditions. In many paintings—for example, Lukas Cranach's *Rest on the Flight to Egypt,* which shows the Holy Family pausing on its way to Egypt—a conifer appears instead of the palm.

The wood wrought from trees is also found in the rituals of many cultures. It has been used for thousands of years in celebrations of the winter solstice—the so-called Yuletide—in northern Europe. As a sacrificial offering under many forms, it was the symbolic equivalent of the living tree. These rituals probably began as elements of a fertility myth. To honor the gods, worshipers set wood alight to scare away the spirits. They believed that at the solstice the sun took up a new life and began its fight against the forces of winter that shrouded everything in darkness. As the days began to lengthen, the sun flaunted its victory over winter. In time, these pre-Christian rituals came to be associated with the Christmas season, in particular the revels of Twelfth Night, on January 5, the eve of Epiphany.

In the Nordic Yuletide festival a whole log was set on fire, the Yule log. Parts of this custom are still found in a number of European countries, from the Iberian Peninsula in the west to Greece in the east. The Transylvanian Saxons in Romania maintained this tradition—also called the Christ wood (*Christholz*), yule clog, or yule block—for an especially long time. In France the wood usually came from a plum or cherry tree, or occasionally from an oak. Before the log was laid in the fire, it was doused with wine or oil, or covered in grains and foods. The residual charcoal was used as a medicinal remedy, while in other countries the ashes were spread over the fields in the holy nights before the winter solstice in the hope of ensuring a rich harvest.

The attraction of all things green, colorful, and glittering in the cold season is elemental. Green has long been considered the color of hope, and midwinter greenery was thought to radiate and summon vitality and fertility, to keep harm at bay. The custom of celebrating the changing year with greenery was already known among the Romans, who used bay branches. In the fourth century, Saint Ephrem the Syrian reported that houses were decorated with wreaths for the festival on January 6. Medieval sources mention evergreen branches, with sharp needles, fastened to the door of the house or hung in the home. Demons, witches, lightning, and disease—they believed—were powerless in the face of this life force.

The solstice evergreen, a freestanding tree usually found next to the village well, was valued in northern and central Europe, reminiscent of the maypole tradition and the raising of a tree to celebrate the harvest. The solstice evergreen was stripped of its bark and branches, except at the top. Sometimes this was later replaced by a new tree-top, and the tree was then decorated with string, small figures, and blown eggs. Christmas and the New Year were both times of transition; they were full of joyful uncertainty and hope for the fertility of the people, the animals, and the fields. The tree had an established place in the life of the community—the girls of the village encircled it, singing and dancing around it. In Sweden such trees were called *Julstänger*. Sometimes these solstice evergreens were decorated with wreaths or rings of greenery.

The winter solstice has long been associated with mythical phenomena, in which ordinary observations are mixed with fantastic visions. In many stories, frostbitten trees and plants began to blossom at the coldest time of year. A text from 1430 includes an account of a "miraculous tree":

> In the harshest and most disagreeable time of year, it bore apple blossoms the size of a thumb on the night of Christ's birth. For this reason many believers from Nuremberg and the surrounding areas would come by and keep vigil in order to see the truth for themselves.

One such winter blossom is the hellebore or Christmas rose, with flowers that are reminiscent of wild roses. In France it is called *rose de Noël* and is said to blossom out from under the snow on Christmas night. Church hymns also recall the flower:

A spotless rose is growing,
Sprung from a tender root,
Of ancient seers' foreshowing,
Of Jesse promised fruit;
Its fairest bud unfolds to light,
Amid the cold, cold winter,
And in the dark midnight.

Christmas was considered a magical time in which the normal rules of everyday life—indeed, even the rules of nature—were suspended. According to one legend, Francis of Assisi went into the garden one winter's night and lay in a thorn bush to taste the suffering of Christ. Subsequently, roses were said to have mysteriously blossomed from the bush. Less mysteriously, in a custom dating to the thirteenth century, cuttings are made of deciduous trees—especially cherry, apple, lilac, plum, hazel, and linden—on December 4, the feast day of Saint Barbara, and brought into the warmth of the house until Christmas so that they may blossom. This practice is prevalent in Franconia, in northern Bavaria. A saying from the region goes, "Are there buds on St. Barbara? Then the blossoms will arrive before Christmas." In the nineteenth century it was common to put a small cherry sapling in a large pot in the corner of the room to await its blossoms.

Light was already an important element of the pre-Christian Germanic winter festival. Remember also the Roman celebration of Saturnalia, conducted solely by candlelight. To this day the Feast of Saint Lucy is celebrated in Norway and Sweden, a festival to honor the martyr whose very name means light. An old superstition states that demons and other harm could be staved off with light during the winter solstice. Candlelight, in particular, evokes the sun. Charlemagne forbade lighted trees because he considered them symbols of hedonistic rituals. But illuminated trees are common in other cultures: consider the holy lighted trees of India and Persia, the Maypole decorated with lights, the old Slavic wedding tree.

Other objects used as tree decorations have long been associated with Christmas. In the tenth century, for instance, it was common in the German lands for children to give apples as presents at Christmas time—a custom that soon broadened to include other foodstuffs such as bread and cheese. In the Middle Ages, Saint Nicholas began bringing nuts, baked goods, toys, and clothing to "good" children on his feast day, December 6; the tradition of giving gifts was considered a Catholic custom, and under the Reformation it moved instead to Christmas.

These traditions anticipate the Christmas tree, but something is still missing. A link can be made between the ritual of our Christmas tree and the paradise play, which had existed since the Middle Ages, even before the nativity play. At a time in which many people couldn't read and books were a valuable possession, biblical stories were dramatized as mystery plays, illustrating doctrinal episodes from creation to damnation to redemption. Creation plays presented Adam and Eve, Satan, and cherubs—guardians of the entrance to the Garden of Eden. Also onstage, of course, was a green tree of paradise, decorated with apples and communion hosts. The paradise play, performed on Christmas Eve, told the story of original sin and the banishment of Adam and Eve from Eden. Human sin was connected to the enjoyment of a fruit—a bright red apple or a pomegranate—and its atonement is set into motion by the birth of Christ. Jesus reaches for the apple that Mary offers him and takes upon himself the sins of the world—a common motif in medieval art. The tree of paradise and knowledge begins to transcend the religious context of the play and move toward a role in the Christmas celebrations of the guilds. Precisely how is not clear, but hints here and there provide clues of the transition.

A Noble Matter

Initially the church saw the Christmas tree as a hedonistic symbol that needed to be repressed. From a purely economic standpoint, moreover, the church had to protect the forested areas it owned. In 1647 the evangelical preacher of the Strasbourg Cathedral, Johann Conrad Dannhauer, lambasted the Christmas tree as a "trifle" and "child's play," which was receiving greater attention "than the word of God and the holy rites."

During the seventeenth and eighteenth centuries, though, the nobility and the wealthy bourgeoisie of the German states discovered the Christmas tree, which quickly became a focus of the seasonal celebration. Embraced by this influential segment of the laity, the tree was finally and fully transformed from a mere symbol of fertility to one of Christianity, competing with the traditional nativity scene. The Christmas tree owed some of its new popularity to the evolution of architecture. These nobles and wealthy burghers were building houses and mansions divided into rooms with discrete functions. A "sitting room" or "parlor," for example, was de rigueur in these new homes—and this provided a place for a tree to be raised and decorated without the children seeing it.

In the small common rooms of the lower classes, there was simply no space for such a tree, even had they been able to afford its increasingly eccentric decorations. But the common people found a way to imitate the bourgeoisie: in the southwest, as well as in Franconia, Thuringia, and Bohemia, it became common to hang a tree from the ceiling joists or rafters. Dangling above, the tree was out of reach of children who might otherwise have pilfered the decorations or sweet tidbits. Trees sometimes were even hung upside down: pointing the root toward heaven was supposed to imbue the tree with divine powers. (Oddly, a form of this custom revived briefly a few years ago in the United States—but with artificial trees!) Hanging the tree from the ceiling like a chandelier was also thought to protect the household from harm. On the other hand, it made placing candles on the tree more difficult. Smaller bushes

A hanging tree from the early nineteenth century

could also be hung in windows. Alas, as these old houses disappeared, so did this custom, because trees couldn't be hung from the new plastered ceilings. The most they could bear was perhaps an Advent wreath or a wooden frame with candles.

Traditionally, Christmas celebrations followed a strict dramaturgy. Having returned from the church service, the father would light the candles. The children weren't allowed in the room until a signal was given or a small bell was rung. It was a moment of heightened emotion, one they had long awaited. Finally the door opened and the little ones were allowed into the room to see the splendor for themselves: the "Miracle Tree," as the German writer E. T. A. Hoffmann called it in his story "The Nutcracker and the Mouse King," and the wrapped presents, which no longer hung on the tree but lay beneath.

The great German poet Johann Wolfgang Goethe, who mentioned the Christmas tree in his writings many times and who is said to have been accused once of illegally felling a spruce tree, wrote in *The Sorrows of Young Werther* "about the joy the little ones would have and of the times when the unexpected opening of the door and the appearance of the marvelous tree with its wax candles, sweets, and apples would put them in heavenly rapture." Sometimes the Christmas story would be read aloud, and then the presents could finally be unwrapped.

The members of the family all gathered around the tree and celebrated together. Worries and arguments were set aside for one evening. The harmonious gathering of the family is an unwritten law. It is a ritual that solidifies the cohesion of the family while also emphasizing the division of the members' roles. It both mirrored and reinforced the domestic order. In some houses a preacher was invited to give a homily or a pianist to provide musical accompani-

Small variants of Christmas trees during the Biedermeier era (1824)

ment. Various illustrators from the nineteenth century have preserved this idyllic family gathering in woodcuts and etchings, and in so doing, they have also cultivated an image of the Christmas celebration that continues to thrive.

In contrast to those who lived in the country, where the forest was only a few steps away, for city dwellers a Christmas tree was something quite exotic. It nourished their desire for a piece of nature, for an open landscape where the wind blew and the air was sweet. Families competed to have the largest and most beautiful tree, the most magnificent decorations, and the greatest number of candles. In other words, the tree became a status symbol. In wealthy homes every member of the family might have his or her own tree, turning the sitting room into a small forest. Thomas Mann describes the spectacle of competitive Christmas tree decoration among wealthy families in his novel *Buddenbrooks: The Decline of a Family* (1901):

> The whole great room was filled with the fragrance of slightly singed evergreen twigs and glowing with light from countless tiny flames. The sky-blue hangings with the white figures on them added to the brilliance. There stood the mighty tree, between the dark red window curtains, towering nearly to the ceiling, decorated with silver tinsel and large white lilies, with a shining angel at the top and the manger at the foot. Its candles twinkled in the general flood of light like far off stars.

An approaching blaze of lights on snowy terrain

But that wasn't enough:

> And a row of tiny trees, also full of stars and hung
> with comfits, stood on the long white table,
> laden with presents, that stretched from the win-
> dow to the doors. All the gas-brackets on the wall
> were lighted too, and thick candles burned in all
> four of the gilded candelabra in the corners of
> the room.

The custom only slowly spread to the lower classes of society.

Who Brought the Tree into the Sitting Room?

Many stories swirl around to explain how the tree gets into the sitting room. Sometimes it is Father Christmas who brings it, sometimes it's Santa Claus, sometimes Saint Nicholas, or just "Father Winter." And this is how it is shown on endless numbers of Christmas cards and paintings. Since the second half of the nineteenth century a flood of such images has saturated Europe and North America. Until that time Santa Claus didn't have any defined look. Now, thanks in part to the work of the German-born American illustrator Thomas Nast, we can see him with a tree on his sled, pulling it with great effort through the high snow. A beautiful American Christmas card from the early twentieth century depicts a less common variant: Santa on a sled with an angel in tow who, in turn, holds a glittering, decorated tree in its arms. In some images, such as one from nineteenth-century Alsace, the tree is already in the room and the *Christkind*—here a woman in a white robe with long, blond hair, a white face, and a crown of golden paper with candles—leads the children to the "tree twinkling with candles." The Christkind was a Protestant invention propagated by Martin Luther in order to take attention away from Saint Nicholas during the European Reformation. Whether in the form of an angel, an androgynous

Nikolaus with tree

small child, or a young woman, the Christkind came at night while all were asleep to bring presents and a tree as well, then escaped unnoticed. The anonymity of this bearer of presents and trees helps to preserve the illusion of Christmas magic from beyond, from a different world. "From out there, from the forest I come . . . ," wrote Theodor Storm.

The old German term for the tree, *Weihnachtsmaien,* was supposedly first used in the town of Schlettstadt in Upper Alsace in 1521. Somewhat later, in 1539, the term *Tannenbaum* (fir tree) appears in a Christmas context in Strasbourg. This word later became popular in northern Germany. But the term *Weihnachtsbaum*—literally, Christmas tree—is first found in 1611, in a felling ban in the Upper Alsatian town of Turckheim. In 1755 a set of regulations for forestry and hunting from Weimar used the term *Christbäumchen* (little Christmas tree), and three decades later in Württemberg one can find the word *Christkindleinsbäume* (the trees of baby Jesus). Weihnachtsbaum and Tannenbaum were common in northern Germany, Christbaum more so in central and southern Germany, as well as in Austria and Switzerland—a difference that remains true today. And with the advent of light decorations in the eighteenth century, the term *Lichterbaum* (tree of lights) was also used for a time.

Christmas trees are usually drawn from the timber stands in the local region. More often than not this means either fir or spruce trees. The German term *Tanne*, from which Tannenbaum derives, was used in common parlance to mean not only firs (the word's literal sense) but also spruces and pines. Pretty confusing. In the case of the fir and the spruce, the confusion is understandable, for at first glance the two genera are relatively similar. But while the cones of the fir stand upright, they hang from a spruce before falling off. Spruce cones fall off in one piece; fir cones disintegrate. If you look closely at the needles of the two species, you will see that the cross-section of the fir needle is flat, while that of the spruce is square. Moreover, fir greens do not prick, but spruces with their pointed needles surely do. Pine trees, meanwhile, with their long needles and irregular build, look completely different and are hard to confuse with other conifers.

Evergreen conifers belong to the oldest order of plants in the world. They have been found in broad swaths of Europe and North America for more than 300 million years. Firs of exceptional age and height have long been held in high regard. Fir trees have traditionally been credited with extraordinary strength and perseverance. And the fir with its little gold pieces, colorful strings, gold paper, flowers, or dyed eggs was thought to protect a new house from lightning and storms.

Botanists have identified numerous species of spruces and firs. Among spruces, for example, there are the warm-weather red spruce, which quickly loses its needles; the blue spruce, which is native to North America; and the Douglas fir, sometimes also called the Douglas pine. For a long time the European silver fir, with its bright green needles, was the classic Christmas tree, but now the stocks have been significantly reduced. Since the middle of the nineteenth century the Nordmann fir has become the most common species of Christmas tree in Europe. The Nordmann fir in fact has its origins not in the far north, as its name would suggest, but in the cool, damp mountainous region on the far side of the Black Sea, in the western Caucasus and northeastern Turkey. This tree, with its dark green, shiny, tight needles, can reach a height of 260 feet. Its even growth also makes it highly desirable. The name comes from the Finnish botanist Alexander von Nordmann, who worked in Odessa in the first half of the nineteenth century. To this day the cones of the tree are harvested— under quite dangerous conditions even for experienced climbers—in order to sell the high-quality seeds to buyers in Europe and North America. According to some stories, harvesters make their job easier by simply lopping off the tops of very old trees. The cones need to be tightly closed before the winged seeds can be broken off and dried.

But not only firs, spruces, and pines became Christmas trees in Europe. The evergreen yew tree, with its poisonous needles and long reputation as a holy tree—in part because it can live for up to two thousand years—should be mentioned here as well. The poet Friedrich W. A. Schmidt wrote in 1795: "The yew stands resplendent with apples full and twinkles with a head of gold and silver." A few years later an anonymous author raved about the "golden fruits" that glitter "in the yew between its swaying branches." But the yew, deterred both by its poisonous needles and the decline of its stocks since the Middle Ages, was never to become a common Christmas tree. The long-leaved evergreen boxwood was also used as a Christmas tree, as documented by Elizabeth Charlotte, Princess Palatinate, in 1711. In medieval times the tree, often found contorted into wonderful forms in Baroque gardens, was also believed to have a magical power to ward off ghosts. Indeed, the Palatinate Christmas tree, which was grown in large wooden tubs, was locally popular for a time. At the Berlin Christmas Market of 1800, "boxwood trees with golden nuts" were offer to all for sale.

The poet Johann Peter Hebel, who in 1803 in his *Alemanic Poems* wrote about the "prickly tree" and "thorny little tree of joy" with "many prickly leaves," is obviously referring to the holly tree, which has not only green, leathery leaves but also natural decoration in the form of its red fruits. This poisonous tree was also hung with little apples and nuts. It is still beloved as a Christmas decoration in North America and Great Britain.

A German document from 1795, the *Simplizianische Wundergeschichts-Calender,* gives an extensive description of a deciduous tree that was erected as a Christkindelsbaum:

It stood there, in the one corner of the room, and its branches were so spread out that they covered almost half of the ceiling of the room and one could stand underneath like a tree in summer. Every branch was covered in all kinds of delicious confectionery and sweets: angels, dolls, animals, and the like all made out of sugar: all of which stood in perfect harmony with the blossoms of the tree. In addition there was gilded fruit of all sorts, in such large quantities that, standing underneath, one felt as if below an arch of edible delights: it was just such a shame that there weren't a few hams or bratwursts and headcheeses, oxen feet, in addition to sautéed grapes hanging from above. In the middle of this storeroom there was the Holy Spirit in its normal form, as the most beloved, beautiful dove of sugar, to the right there hung the infant Jesus, and to the left his mother—a very sweet sight it was, everything made of sugar, such that I could have happily gobbled them both up, the Virgin Mary and her child, had that been allowed. Finally, the entire tree itself, with all its branches and fruits, was covered with a golden net made of many thousands of gilded hazelnuts hung in a row and decorated with tinsel and rings. Between all these delicacies an uncountable number of wax candles twinkled forth, like stars in the sky, completing the magnificent sight.

In the end deciduous trees did not manage to make a place for themselves, most likely because of the difficulty of getting them to turn green in winter.

Alternatives to Wood: Pyramids and Klausenbäume

Beginning in the eighteenth century a pyramid constructed out of wood and bearing candles, decorations, and presents came into fashion. The origins of this object are obscure, whether based on the Christmas tree or independent of that symbol. There are hints, though, that its predecessor was greenery from evergreens that was tied together into the form of a pyramid. The pyramid, decorated and fitted with candles, evokes a lighted tree. It is found in the early Christmas services of evangelical churches and community celebrations before making its way into the home. Moreover, in the context of early medieval churches it is related to the use of *arbores*—candelabra in the shape of a tree, consisting of winter greens tied together and presented as the tree of paradise. So, too, is the menorah, the nine-armed tiered candelabrum used in the Jewish tradition, reminiscent in its shape of the tree of life. The artisanal centers of pyramid production were in the Erzgebirge (the Ore Mountains region) and Vogtland in eastern Germany. The old Bavarian *Klausenbaum,* which is erected during the Festival of Saint Nicholas, is a stylized one comprising three wooden branches, stripped of their bark, decorated with boxwood greens, and leaned together to form a peak with an apple at the top.

For all their craftsmanship, because the pyramids could be used again and again, they remained a substitute Christmas tree for those without the means to buy the real thing every year. As the population and prosperity grew, so too did the demand for Christmas trees. This led to the cultivation of trees especially for this purpose, which, as their supply increased, became less expensive.

The "Fir Tree Religion"

For a long time the Christmas tree in Germany was considered Protestant—a *Lutherbaum* (Luther tree)—and the aversion of many Catholics went so far that at the end of the nineteenth century many simply called Protestantism the "Tannenbaum religion." The custom therefore became common earlier in the evangelical regions of northern Germany than in the Catholic parts of the south and west. In the middle of the nineteenth century Christmas trees were already found in churches—decorated with angels, crosses, wafers, and straw stars. These last were meant to represent the Star of Bethlehem while also reminding the viewers of the straw in the manger.

The folklorist Dietz-Rüdiger Moser has written,

Handicrafts around the Christmas tree (1883)

Luther and his followers fought for the traditional manger celebrations and customs with their scripture-based understanding of the celebration of Christ's birth, above all the associated shepherds' play, so that in the end a completely different Christmas celebration came about: with readings of the Christmas story at home, with the singing of Christmas songs, with certain gift-giving customs, and so on.

Thus, according to Moser, the tree decorated with candles asserted itself in the culture in a way that was impossible in the context of the Catholic manger celebration. Although the reservations of the Catholics gradually lessened, as late as 1909 the Benedictine monks Augustin Scherer and Johann Baptist Lambert lambasted in their *Lexicon for Preachers and Catechists* the "fraud" of the Tannenbaum tradition, citing Christmas trees erected for cats and dogs, even trees erected atop graves.

Tree Decorations through the Ages

As the Christmas tree came within reach of less privileged parts of society, the nature of its decorations changed. Until the nineteenth century mostly edible objects were used: baked goods, sweets, apples, and nuts. Candies were considered precious before the time of mass production, because of the time and skill needed to make them, and children were not alone in being tempted to secretly pilfer and nibble tidbits from the tree. In 1793 Friedrich Schiller was found to have succumbed to the enticement of the "tree of edible delights." The doctor Friedrich Wilhelm von Hoeven wrote, "On Christmas Eve I visited him and what did I see? An enormous tree, lit by countless candles, with gilded nuts, gingerbread, and all sorts of little sweeties. In front of it Schiller sat all by himself eating of its fruits . . ."

Enduring yuletide confections include sugar bread and other baked goods made from egg dough flavored with anise seed. *Springerle,* for example, are formed into small squares with a motif on them, such as a knight or birds, but they are also sometimes made to look like hearts, curls, flowers, or pretzels. These motifs are made by pressing the dough into a wooden mold and carefully removing it and dusting it with flour before baking. Since the cookies are baked at a low temperature, the motif remains easily visible. When cooked, the Springerle are painted and hung on the Christmas tree by a woolen thread. The thin *Tirggel* biscuits of Switzerland are subjected briefly to high heat, which brings out the figures, script, and ornamentation molded into them; these are also painted after baking. A different process is required for dark gingerbread, the contours of which would not stand high heat. These are decorated instead with light-colored icing or wafers bearing small figures, creating a pleasant contrast. Later, colorful images embossed on paper were used instead of wafers, and the cookies thus created became popular collector's items in Germany.

Edible figurines are made by melting white sugar and cooking it until it forms a thick mass, then placing it in a form and folding the form so that the figure takes on an even shape. More durable than these almost transparent figurines were those made with gum tragacanth. Tragacanth is a resin made from boxthorn that grows in India and Iran. It found its way through Italy to central Europe and was first used there as a binding agent by apothecaries. Mixing the thick juice of the plant with sugar, egg white, starch, and rose water produces a pliable, moldable dough that can be pressed into a form and dried in the oven. Tragacanth—often described as "cheap candy"—is distinguished by its malleability and shelf life, but since it does not taste as good as marzipan, it has largely been forgotten today.

At times small fruits, turnips, and pretzels were hung on the tree in little baskets. Theodor Storm provides an account of "squirrels made of marzipan in half-size, with a raised tail and clever eyes," and a "rabbit with a leaf of cabbage in its front paw." Little sacks were also filled with candies, and candy canes were wrapped in metal spirals.

Beginning in the nineteenth century Christian symbols became increasingly common on the Christmas tree. Figures fashioned of colored wax became popular: the infant Jesus or angels bearing swaddling clothes. These were decorated with wool, wire, paper, and glass silk. Some figurines were made entirely of wax; others were made by pouring wax over a papier-mâché core. Either way, they were quite sensitive to heat, and it was not uncommon for an arm to melt off. In the nineteenth century Christmas tree decoration reached a new level of refinement with the shaping of molten glass: blown, mirrored, and colorfully painted birds, birds' nests, frosted glass spheres with little decals, and colored eggs were made from this medium, as well as lanterns made of tin, various paper animals, and magnificent carriages made of Dresden paper.

Tinsel was probably inspired by the so-called *leonische Drähte* (introduced by Huguenots from Lyon)—silver or gold-plated copper wire that was originally a by-product of metal work. It is reminiscent of the silver thread that was woven into church vestments in the Middle Ages. For a long time tinsel—also called "silver-plated sauerkraut" in colloquial German—was cut from tin foil. It is reminiscent of a thin icicle, but it could just as well bring forth summery associations. In 1884 Theodor Storm wrote, "On the Sunday before Christmas my friend Petersen brought a sack filled with marvelous silver thread. The tree wrapped in this fine silver thread looked like a flying summer." A variation was the so-called angel's hair, fairy's hair, or baby Jesus' hair—fine metal thread that also helped to give the tree a glimmering beauty.

In the country many forwent elaborate and expensive tree decorations. Beginning in the 1870s American Christmas trees displayed so-called *Dresdens* (named after the Saxonian city): three-dimensional paper ornaments that were exported to toy wholesalers like the Erlich Brothers in the United States. Two pieces of damp cardboard were stamped and given shape, glued together, and lacquered. They were shaped like hearts, bears, mice, violins, or just little boxes, and candy was placed inside before the Dresden was hung on the tree.

At the end of the 1870s there is documentation from the Carinthian Gail Valley in southern Austria that a thick spruce, free of all decoration, was placed in the corner of a farmstead as a sign of silent joy. On frosty cold winter mornings the tree, now covered with little icicles and illuminated by the sun's rays, shimmered like a Christmas tree covered in lights—without any tinsel or fairy's hair. The wild beauty of the tree sufficed.

Rosy-Cheeked Apples Made of Glass

Glass spheres as light as a feather—a ball with a diameter of three inches weighs less than half an ounce—are the most prominent trimming on today's Christmas tree. As light reflects on their surface, they twinkle and glisten. The red spheres are reminiscent of an apple or a pomegranate, both of which have been symbols of fertility and immortality since antiquity, only later repurposed in the context of original sin. The golden balls could be seen as miniature likenesses of the stars in heaven.

It is possible that the origin of these ornaments can be found in the Thuringian forest. There the craft of blown glass can be traced to the beginning of the seventeenth century, when immigrants from Bohemia built their first

The tree as romantic atmosphere

huts. According to a story from the Lorraine region of France—another place with a long glass-blowing tradition —the first Christmas spheres were produced in a village there called Meisenthal. A small variety of apple is cultivated in Alsace, Christ's apple (*Christapfel*) or ice apple (*Eisapfel*), and this fruit traditionally was used to decorate the Christmas tree. In 1858 a drought caused the harvest to be lost, so glassblowers made red glass spheres that could be used on the Christmas tree instead of the apples.

To make a simple ornament, a prepared glass tube is heated over a flame and blown into a sphere. More ambitious craftsmen make strawberries, acorns, and nuts of glass, inspired by fruits that traditionally were used to decorate hats. Owls and bells were also made, and even necklaces of glass pearls. Great amounts of wood were needed to fire the ovens, until 1867, when modern gas ovens began to be used in the Thuringian city of Lauscha; with the gas oven a much hotter flame could be generated, and larger numbers of spheres and figurines could be produced. Initially the spheres were given an inside mirror coating of a tin-and-lead alloy, lending them a silver-gray sheen. In 1870 Justus von Liebig developed a coating of silver nitrate, which provided the same visual effect with a much lower health risk. The glass balls became an international export item. By 1900 Franklin W. Woolworth was already importing them to America. To this day the traditional, handmade balls are highly valued, especially those that look as if their surface is covered with tiny ice crystals. Children especially love seeing the distorted reflection of their faces in the balls. Is a Christmas tree without glass balls like the night's sky without stars? And the history of these decorations would not be complete without noting that today these once-fragile balls are often made of plastic; thus is a little romance traded for durability.

At the Top of the Tree

For a long time the very top of the tree was not accentuated at all (this being impossible, of course, with the hanging trees). Only in the nineteenth century did people begin to put an ornament on the top. It might have been a proud Angel of the Annunciation dressed in a skirt, a knight made of beaten egg whites, a bird of paradise made of golden paper, a small rooster, a glockenspiel, a rosette made of gold foil, brass, or glass, or simply a golden apple. Sometimes a star with a comet's tail adorned the treetop. This evoked the Star of Bethlehem, which guided the three magi from the East. The production of the angel—at first, made of wax or papier-mâché, later of biscuit china, with an expression that was sometimes quite stern and other times more friendly—was closely tied to the dollmaking industry in the towns of Nuremberg and Sonneberg in south-central Germany. Sometimes the angel wore a banderole bearing a legend, such as "Peace on earth to men." By the end of the nineteenth century such figurines had been supplanted by elaborate glass peaks, reminiscent of the Prussian spiked helmets. In the twentieth century fashion turned to a device with chimes that, when heated by the warmth of the candles, was set in motion and made a sound—if you believe the euphoric descriptions of the manufacturers—that gave the illusion of church bells ringing in the distance. Various such chimes were available, but the typical design was a small brass angel blowing a horn whose movements caused the gentle ringing of the other metal elements. A further variation was the *Viererglocke,* an artful device in which a spring-powered tin Saint Nicholas lifted his arm and pulled a cord to set four bells ringing.

Fascination and Danger: Trees Full of Candles

Did candles on the Christmas tree start the blaze that destroyed the Castle Warthausen in Upper Swabia in 1621? Probably not, for many indications suggest that decorating trees with candles did not become widespread until the following century. Did Martin Luther, enchanted by the stars in the sky on Christmas Eve, light the first candle on a tree? That too is probably but a legend. Elizabeth Charlotte, Princess Palatinate, wrote in 1708 of "little candles" affixed to a boxwood branch: "It looks adorable." Royal decrees later banned the use of candles in order to reduce the risk of fires. Goethe saw a Christmas tree with candles in 1765 in Leipzig and is said to have introduced the fashion to the Weimar court a decade later. In 1821, on the occasion of the Christmas celebration of Karl August, Grand Duke of Saxe-Weimar-Eisenach, Goethe wrote:

Luminous trees, brilliant trees,
From head to toe in sweets,
Swaying in the light.
Moving hearts young and old,
Such a festival to us is given;
The beauty of the gifts adored,
With wonder gaze we up and down,
Here and there and once again.

In a way the lights on the tree are an apt symbol of the birth of Christ, who was himself described as the "light in the darkness." Before the introduction of electricity, light was treasured. A tree decked with candles created a completely new atmosphere; its radiance was much greater in the darkness of the past, before neon lights and lighting of all sorts on the street and in our homes dimmed its comparative luster. The characteristic scent of beeswax candles also contributed to the atmosphere, but they were very expensive. More common were mass produced tallow candles and oil lamps. One could fill one half of a nutshell with a little oil, dip a wick into it, and fasten it to the branch with string or wooden clips. Small glass lamps were also filled with oil; a small pin connected to the lamp could then be affixed to the branch of the tree. These lamps, however, produced a lot of soot. Once beeswax candles were replaced by stearin, made since 1818 of plant and animal fats, and paraffin, processed from crude oil after 1830, candles became affordable to the broader population.

Fastening the candles to the tree required great skill

and care. Before the invention of special candleholders with clips, made of bronzed and painted tin, the tree trimmer had to make do with pins. After cutting off the head of the pin with a pair of pliers, one heated the dull end and inserted it into the bottom of the candle. Then the candle could be pinned onto a branch. But this sounds easier than it actually was, and it was not uncommon for a tree to go up in flames if a single candle snapped off its branch. The metal candleholder clips—described in a brochure as "candleholders to clip on or balance on the branches, white, gold, and various colors, starting at just ten cents"—were often made to resemble butterflies, angels, and birds. These candleholders didn't eliminate the risks, but they did attach the candles to the tree more securely and safely. A small hanging counterweight, in the form of a tree or an icicle, ensured that the holder remained balanced on the branch. Other candleholders were screwed directly into the trunk. One variation involved lights that could be hung like a small lantern, with red or green translucent windows. The lighting of the candles became a little performance in itself. Traditionally, an even number of candles was always lit, starting at the top. This was done with a tubelike wax candle, up to three feet long, that could be extended to help both light the candles and extinguish them (this time from the bottom up). We have no reliable fire statistics from the time, but from time to time a Christmas tree fire made news when a person was injured or a house burned down.

The search for safer tree illumination was ongoing. In the United States in the 1870s one inventor created a gaslit cast-iron Christmas tree with simulated branches that featured little openings where the gas could be ignited. "The Improved German Christmas Tree," as it was called, was exported to Germany as well. It was distinguished not only by its little blue flames but also by its clearly audible hissing: "The gas flows through the hollow branches and, where a candle once stood, there now flickers a gas lamp from the narrow slit." One can only assume that this mechanical wonder led to its share of accidents. At any rate, this temporary solution failed to establish itself.

Around 1885 gas lamps were replaced by the electric lights we know today. Some claim that Berlin was home to the first electrically illuminated tree. The wires—wrapped in green—that con-

Advertisement for a chiming Christmas tree

nected the lights were hidden in the tree as well as possible in order to maintain the illusion that they were "real" candles. And the little lights took the forms of glass fruits, Santa Clauses, suns, and moons, all lit from inside. East Asian companies initially thrived in this market.

But did the electric lights create the same atmosphere as the real candles? Some traditionalists remained skeptics, but many used the new lights lest they be seen as "old fashioned." In the end there was one overwhelming argument: the risk was gone.

Solid Footing: A Diversity of Christmas Tree Stands

How do you get the tree to stand up securely? One early stand was a wooden cross, painted green or covered with moss or stones, into the middle of which a hole had been drilled. Another option was a stool with a hole in which the tree was wedged. Sometimes the tree was placed in a tub of water or a bucket with wet sand, thus keeping the needles green for a few days more. In the 1860s cast-iron stands became more common, shaped to resemble gnarled roots, angels' heads, Santa Clauses, or pine cones. These ornate stands became gradually simpler in the twentieth century, often decorated with simple geometric designs. The basic principle remained a stand with three or four legs, with screws at the sides and a spike at the bottom to center and secure the tree. One variant incorporated a group of movable birds or gnomes made of sheet metal on a pipe that

An unusual Christmas tree stand

accepted the trunk. Once the tree was inserted, the beaks of the birds or the peaks of the gnomes' hats screwed into the trunk to secure it. Some stands, usually made out of hoop iron, had hinges and could be folded up or taken apart. Others could be filled with water. The description from the Dresden company E. Neumann & Sohn from the year 1889 reads: "Place the tree as deeply as possible in the water and secure it with the screws. If the water is frequently replenished and mixed with a small amount of powdered charcoal, then the scent of the needles will keep longer and prevent them falling off all together." One especially elaborate mechanical Christmas tree stand from 1893 featured a hand-wound clock movement. The tree rotated around its axis accompanied by the melody of "Silent Night." The advertisement promised a "singing, chiming little tree." More expensive models included a music box in a separate walnut, which featured up to eight melodies on a roller or, later, a metal record.

In times of adversity imagination knew no bound-

Hans Christian Andersen and his tree

aries. Some wedged the trunk into the hub of a cartwheel or cut a rutabaga in half and drilled a hole to accommodate the tree. Cast-iron stands have now been largely replaced by those made of ceramic and plastic.

The Personalized Christmas Tree

From time to time the Christmas tree has been given almost human attributes. The German poet Rainer Maria Rilke wrote in his beautiful poem "Advent" that a Christmas tree anticipated "how soon it would turn pious and holy." Even more vividly, though, the Danish writer Hans Christian Andersen personalized the tree in a story from the year 1844. In "The Fir Tree" he describes a tree that is eager to grow tall and fulfill its fate. In its third year the tree is finally so tall "that the rabbits had to go around it." It watches as the lumberjacks come and fell the largest trees. In the spring a stork tells the tree that those trunks were used as boat masts; the bird had seen it in Egypt. At first that inspires the tree's desire to sail over the seas, until a sparrow tells the tree that some of the smaller trees had a different fate: "They are planted in the middle of a room and decorated with the most wonderful things, golden apples, honey cakes, toys, and many colorful lights." Now the tree wants nothing else than to take "this brilliant path." When it is finally cut down, it feels a terrible pain and falls unconscious. It awakes to find itself in a magnificent hall, decorated with paintings and Chinese vases. It is placed in a cask filled with sand.

> A servant and madam circled him and decorated him with little nets of colorful paper, each strung with confections; gilded nuts and apples were hung and over a hundred blue, red, and white candles were placed on the branches. Small dolls floated among the greens and high upon the tip sparkled a star made of gold foil. It was unbelievably glorious! The tree could hardly await the evening when the candles were finally lit. Several children came into the room and began to take the little presents off of the branches. They ravaged the tree until "all the branches were snapped."

The tree happily anticipates being decorated again the following morning, but instead it is placed in a dark corner by a servant boy and girl, only to be hauled into the courtyard a few days later. "The children, who had danced around the tree on Christmas Eve, came by and called, 'Look what's still hanging on that wretched old tree!' And they stepped on its branches until they bent and broke." Finally the tree is cut up into pieces and burned. "Christmastime was now over and so was the tree and the story as well; gone, gone, and so it is with all stories!" sounded the grim conclusion of Andersen's story.

The Europeanized Christmas Tree

Since the nineteenth century the Christmas tree and everything connected with it has gradually expanded from central Europe across the entire globe. It is a modest cultural transfer that is not written about in the tomes of history, but hints are found here and there. Popular legend has it that the custom spread to England after Queen Victoria and Prince Albert wed in 1840. Queen Victoria then had a forty-two-foot-high tree decorated with valuables and gifts, for a cost of £10,000. However, brief notes dating back to an earlier time suggest that the tree's voyage to England had different roots; it might even be seen as a joint effort. Samuel Taylor Coleridge observed the Christmas tradition in a German home and described it in a letter he sent in 1798:

On the evening before Christmas day, one of the parlors is lighted up by the children, into which the parents must not go; a great yew bow is fastened on the table at a distance from the wall, a multitude of tapers are fixed to the bough, but not so as to burn it till they are nearly consumed, and colored paper, etc., hangs and flutters from the twigs. Under this bough the children lay out in great order the presents they mean to give their parents, still concealing what they intend to give to each other. Then the parents are introduced, and each presents his little gift; they then bring out the remainder one by one, from their pockets and present them with kisses and embraces. I was very much affected. The shadow of the bough and its appendages on the wall and arching over the ceiling made a pretty picture.

Of course, knowing of a custom does not automatically mean adapting it. It is unknown whether Coleridge's report made any impact.

Even half a century later the Christmas tree still seems to have been perceived as something new and foreign. In 1850 Charles Dickens, who is himself forever associated with Christmas, offered a unique interpretation of the tree, describing it at the time as a "pretty German toy." After giving a vivid description of the tree with its "rosy-cheeked dolls, hiding behind green leaves" and "broad-faced little men, much more agreeable in appearance than many real men," and citing a child saying, "There was everything, and more," Dickens concludes with an almost religious undertone:

> This motley collection of odd objects, clustering on the tree like magic fruit, and flashing back the bright looks directed towards it from every side— some of the diamond eyes admiring it were hardly on a level with the table, and a few were languishing in timid wonder on the bosoms of pretty mothers, aunts, and nurses—made lively realization of the fancies of childhood; and set me thinking how all the trees that grow and all the things that come into existence on the earth, have their wild adornments at that well remembered time.

Even before the middle of the nineteenth century Christmas trees could be found in Italy, in the Netherlands, and among the upper classes of tsarist Russia. In 1837 the duchess Hélène of Orléans is said to have had a tree erected in the Tuileries Palace. A certain reluctance on the part of the French middle class, who considered the tree "an intruder of Alsatian origin," prevented this custom from being widely popular before the 1870s.

Each region, of course, brings its own sensibilities to the tree, sometimes resulting in a practice innovative or unique. In Scandinavia it became a popular custom to perform a round dance and sing around the Christmas tree. Even the decorations can vary. In Norway—where the Christmas tree, or juletreet, was once considered a "sinful and blasphemous" form of "idolatry"—it is popular to hang *Nisse,* little gnomes with red hats, from the branches. Norway has also been the source, every year since 1947, for the fifty-foot-tall, fifty- to sixty-year-old fir erected annually on Trafalgar Square in London.

American Trees

The American way of celebrating Christmas is relatively recent, evolving in the latter part of the nineteenth century. In the United States the roots of the Christmas tree reach back to the early 1800s—perhaps not as far as we might guess, given the ubiquity of German immigrants in early America. Few seem to have brought the custom along, which in turn suggests limited diffusion of the tree in their homeland. In any case, such an icon surely would have been reviled by the Puritans, who associated evergreens with heathen superstition. In fact, even Christmas itself was denounced as an artificial invention without any biblical warrant. As the American historian William B. Waits has unequivocally stated: "Good New England grandmas through the first half of the nineteenth century disliked Christmas. The public sector was in accord; for example, December 25 was not a school holiday in the antebellum Northeast and children were required to attend classes."

Christmas scene, drawn by Lewis Krimmel

Calvinist reservations prevailed, but a different reality was evolving under the surface. As early as 1812 John Lewis Krimmel from Germantown, Pennsylvania, created a drawing of a rather small illuminated tree placed on a table and surrounded by a miniature family and its dog, all of which was enclosed by a picket fence. And in 1823 the Society of Bachelors in York, Pennsylvania, promised that the decorations on their *Krischtkintle Baum* would "be superb, superfine, superfrostical, shnockagastical, double refined, mill'twill'd made of Dog's Wool, Swingling Tow, and Posnum fur; which cannot fail to gratify taste."

Documents confirm that a Christmas tree was found in the house of Charles (or Karl) Follen, a literature professor at Harvard University of German descent, in the year 1832. It had "7 dozen wax tapers, gilded egg cups, paper cornucopiae filled with comfits, lozenges and barley sugar." And Daniel Pool writes in his little volume on the Christmas tradition in New York, "In the 1830s New Yorkers were venturing across the East River to the (then separate) municipality of Brooklyn to view that city's Teutonic population's 'custom of dressing a Christmas tree.'"

Still, this documentation of the existence of the tree in the early nineteenth century should not be misread, for Christmas trees remained exotica for some time, eyed with both interest and skepticism. Again and again travelers to Germany sent back reports to American newspapers of their fascination with "installation of the Christmas tree." As hard as it is to imagine today, Christmas and Christmas trees still struggled to find their place in America—sometimes, indeed, still drew pronounced hostility. As the progressive writer Lydia Maria Child observed in 1845, "The Puritan blood still flows too briskly in my veins to allow me to relish over much the Christmas tree."

But if in the public perception at the time the tree was clearly identified with its German origins, that was not always cause for disquiet; sometimes it inspired explicit approval. In January 1856 *The Living Age* printed a poem signed E.G.B. that included the following stanza:

Of all the store our German brothers brought us,
Long since, across the sea,
Best do I love the simple faith they taught us,
Linked with the Christmas tree.

An observer in the *Philadelphia Bulletin* wrote in 1860 that in Pennsylvania, "where the English and German sentiments both survive . . . there is scarcely a house that has not its Christmas tree."

In December 1883 the *New York Times* noted that "comparatively few Christmas trees were exposed for sale this year" and went on with a rather harsh judgment:

> The German Christmas tree—a rootless and life-less corpse—was never worthy of the day, and no one can say how far the spirit of rationalism which begins with the denial of Santa Claus, the supernatural filler of stockings, and ends with the denial of all things supernatural, has been fos-tered by the German Christmas trees, which have been adopted so widely in this country.

The author pulled even more stops to set the readers against the tree:

> The Christmas tree, dropping melted wax upon the carpet, filling all nervous people with a dread of fire; banishing the juvenile delight of opening the well-filled stocking in the dim morning light,

and diffusing the poison of rationalism thinly disguised as the perfume of hemlock, should have no place in our beloved land. It has had its day, and the glorious reaction in favor of the sacred stocking will sweep it away forever.

As we know, this prophecy didn't come true, but the tradition of the trees was adopted only gradually, not gaining popularity in wider circles until the second half of the nineteenth century. Sometimes trees were introduced in Sunday schools to encourage attendance. As Karal Ann Marling has noted, "Some efforts were made to find a religious justification for decorating them," which, of course, was difficult or impossible, given the symbol's clouded origin. In the 1850s *Godey's* magazine wrote about an "orthodox Christmas-tree," which would "have the figures of our first parents at its foot, and the serpent twining himself round its stem."

Ultimately, Calvinist attitudes took a back seat, as the tree entered the national imagination and even became the icon of American Christmas, just as it had done in Europe. In a sense, the Americanization of the "German" Christmas tree runs parallel to the Americanization of German immigrants.

In 1880 no fewer than 200,000 trees were brought from the Catskill Mountains and the whole New England region by train, boat, and wagon to New York City's Washington Market, a popular place for buying wholesale foods and vegetables. Alongside turkeys and Santas, heaps of trees signaled the coming Christmas season. *New England* magazine reported in 1895 how "eagerly customers flocked to purchase the mountain novelties, at what appeared to the unsophisticated country man very exorbitant prices." And what was purely a men's affair before, going into the forest in search of a suitable tree, came more into the women's sphere. They could now take part in the selection of a tree.

Although the North American forests seemed to offer an almost unlimited source of trees, at times in the late nineteenth century trees in the vicinity of cities were scarce. Instead, newspapers suggested using the branches of broadleaf trees and decorating them with wool. And these trees became popular.

Christmas trees in America have a number of distinct characteristics. Trees with especially dense growth are always most sought after. In order to spur on the growth of more branches, the trees are meticulously pruned, with a focus on the thinner, lower branches. In this way, the distance between the branches of the tree can be reduced, thus increasing its aesthetic value. The disadvantage is that these trees are dangerous for wax candles because they present a high risk of fire. Some American tree gardeners do not scruple even to spray-paint trees in order to maintain their fresh appearance after they have been felled.

Decorating the Christmas tree, too, can be more expansive in the United States than is traditional in Europe. For some, the flourish may be an oversized snowflake made of bast, miniature facades of famous buildings, or colorful ribbons; for others it might be lamps, tinsel garlands, and long white bands hung in the branches spiraling up to the ceiling. Almost anything is possible, and often it seems—at least from a European perspective—that the motto is "the flashier, the better."

But generalizations are dangerous. Some trees do not bear fanciful decorations, but, rather, are covered from head to toe in lights, such that in the dark the tree's form is clearly visible. This heavy use of lights—often in the yard in front of the house—is a cliché of the American Christmas celebration—and of American Christmas films. Symmetrical trees are highly regarded as well—a preference famously lampooned in the popular television special *A Charlie Brown Christmas,* which encouraged affection for imperfect trees. A "Charlie Brown Christmas tree" can be anything from a skimpy tree with but a few branches to a single spruce branch decorated with one red ball.

In the Blue Room of the White House a meticulously chosen tree is raised every year. Traditionally this is the responsibility of the First Lady. Only President Theodore Roosevelt, well known as a conservationist, at a time when few trees were planted specifically for this market, forbade a Christmas tree in the White House. Once, two of Roosevelt's sons felled a small tree and placed it in their room. When the president learned of this he took the boys to a forester friend of his, who did not admonish them as expected but recommended that, next time, they should

Jacqueline and John F. Kennedy next to the Nutcracker tree

dig out the entire tree by its roots and then take proper
care of it. That year the family did indeed celebrate the
holiday with the small tree.

The selection of the tree for the president's family is
by competition. The winner is allowed to present his treasured tree to the First Lady. Jacqueline Kennedy's tree, inspired by Tchaikovsky's *Nutcracker Suite* ballet, is a White
House legend. She decorated the tree with toy soldiers

and instruments, lollipops, miniature fruit baskets, fairies, and candy canes. Even more famous than the tree in the White House is the National Christmas Tree, which has been erected near the White House every year since 1923. The Society for Electrical Development sponsored it, well aware that it could promote outdoor lighting at Christmastime. Soon after, loudspeakers were installed in the tree, from which Christmas songs were played; thus was born the "Singing Tree."

The Christmas tree is so beloved in the United States that one finds it again, in a somewhat different guise, just three months later in the form of the Mardi Gras tree. The decorations for this pre-Lenten version are louder, and thus the tree is a touch less elegant than the Christmas tree itself. Colors such as yellow and violet dominate, and masks, large feathers, and colorful strings of beads decorate the branches. The tree comes down on Ash Wednesday.

Our brief history of the American Christmas tree would not be complete without mentioning that some American Christians believe to this day that the Christmas tree is a pagan and sinful custom. They are disturbed by the place of honor and attention the tree receives during the holiday season. Some base their stance on the Book of Jeremiah, chapter 10, which warns, "Do not learn the ways of the nations. . . . For the customs of the peoples are worthless; they cut a tree out of the forest, and a craftsman shapes it with his chisel. They adorn it with silver and gold; they fasten it with hammer and nails so it will

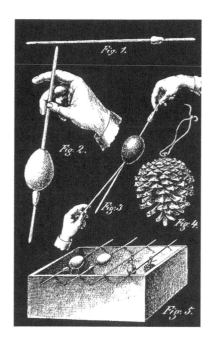

An art unto itself:
gilding tree decorations

not totter." The admonition extends to humanity: "Like a scarecrow in a cucumber field are they, and they cannot speak; they must be carried, because they cannot walk!" Defenders of the Christmas tree would argue, though, that the tree Jeremiah refers to is an idol, and the decorated Christmas tree is not. If it were, would its worshippers simply throw it away at the end of the Christmas season? Most Christians enjoy the tree with a clear conscience.

Fashions of Christmas Trees

Finally, by the late nineteenth and early twentieth centuries, the Christmas tree had become the central focus of Christmas Eve in family life, now in all social classes, although in some regions—the remote valleys of the Alps, for example, and even in the Swiss countryside—the tradition of the Christmas tree gained no foothold.

Some specialists transformed decoration of the Christmas tree into an exceptional skill. Among them was the German Hugo Elm, who, in his 1878 *Golden Christmas Book,* made a plea for "a tasteful separation of the numerous decorations on the tree" in order to avoid a "bland hodgepodge." He suggested the following steps, precisely designed for the anatomy of the tree and the load capacity of its branches:

> Decoration should begin with the heaviest objects, which are best placed near the trunk and in the middle of a branch. Next one should place the nuts. Place silver and gold nuts alternating— about 3–4 pieces on the longer and 2–3 on the shorter branches and on the top, smallest branches only one each. The golden and silver pinecones, in contrast, should be placed farther forward in the second third of the branch, as calculated from the trunk outward. Marzipan and sweets are best placed in between two nuts. Shiny glass balls, fruits, and the like are to be placed preferably on the upper branches in order to enjoy the effect of their refracting rays of light. Metal coils and tinsel are spread out at the tips of the secondary branches, for these are thinner and are more likely to sway than the thicker main

branches. The small baskets and nets made of paper are placed on secondary branches. The individual stars should be distributed evenly while the strings of alternating nuts, straw, stars, paper, and similar are to be wound around the branches and distributed. Paper bags should always be put on the tips of the branches, ideally beneath the lights. At the top of the tree one customarily puts a large star made of cardboard covered with golden paper, in which one glues either a self-made or bought Christmas angel. A thick tome with golden fringe and an old Gothic script displaying the sublime Christmas saying "Glory to God in the high" also looks magnificent. Once the lights have been put on the tree, the tops of the branches can be covered with loosely pulled cotton and these then affixed with silver thread.

During the second half of the nineteenth century fresh-cut Christmas trees, giving off their fragrant scent, were sold complete with decorations in Christmas markets in Dresden, Frankfurt am Main, Berlin, Nuremberg, and Vienna, and from then on the practice became more widespread. The English writer Frances Trollope spent her Christmas of 1837 in Vienna and reported,

At the corner of every street we see customers of quite the lower orders bargaining for trees, adorned with knots of many-colored paper, in order to celebrate the Christmas. These trees . . . are provided of every variety of degree, as to size and expense, by nearly every family in Vienna where there are young people. . . . The tree is called "the little tree of Jesus"; and on its branches are suspended all sorts of pretty toys, bijous, and bon-bons, to be distributed among those who are present at the fête.

A tree at the Christmas bazaar in Hamburg in 1853 reportedly rotated in the middle of an enormous carousel filled with mythical creatures.

The opulent trees of the late nineteenth century, such as the one described by Hugo Elm, were too kitschy for the progressive spirit of the twentieth century. Concentration on the "essentials" was called for. But what did that really mean? Silver blossoms, tinsel, icicles, cotton snow, and white candles were meant to emphasize the wintery picture—a veritable "snow tree." Flat glass silk, for example, was used for birds' tails, on paper stars, and for the delicate wings of butterflies, creating a "white wave." Other trees, meanwhile, were decorated with small exotic glass animals such as snakes, fish, and crocodiles, or with fanciful objects like musical instruments, crowns, or stars, sometimes sprinkled with glitter or "Venetian dew." As varied as these art nouveau trees could be, their filigree decorations gave them a subtler, finer effect.

Thick books could be filled with the countless little

And they even found the perfect tree

things that have hung from Christmas trees, today and in times past. Since most of the expensive and intricate objects could be reused—if properly stored in boxes—decorations were often handed down from one generation to the next. Innovation, therefore, occurred only slowly. One clear example of such tradition is demonstrated by the Berlin couple Anna and Richard Wagner. For many decades, beginning at the turn of the twentieth century and continuing until the time of the Second World War, they took a photo of themselves and of their tree every year. They got older, but the trees and their decorations stayed the same, almost down to the branch.

In harder times, such as during the wars, many had to celebrate Christmas without a tree. A few pieces of greenery, a candle, or a burning pine shaving served as reminders. The solutions to this vacuum were varied, imaginative, and original. Some drilled holes in a broomstick and stuck in fir sprigs—thus was born the "fake" Christmas tree. In areas with few forests, such as on the islands of the North Sea, one had to make do with such substitutes.

At such times of crisis Christmas trees also became unaffordable for some, a prohibitively expensive wish for those less well off—and a painful reminder of poverty. These poor souls might be reduced to pressing their noses against store windows displaying Christmas trees—and hoping for the opportunity to purchase one at the last minute or even after Christmas for a reduced sum from a seller who failed to sell all his trees. Parents consoled their children, claiming that Santa Claus was late this year—and the children, perhaps, believed them.

Children of poor families were also able to admire decorated trees in churches and schools; sometimes they were allowed to help take off the decorations. And in other cases, when children importuned too loudly for a tree, one father or another might make his way into the forest to get—or steal—a tree. The German author Wolf-dietrich Schnurre, in his short, touching story "The Loan," tells of a father and son who procure a blue spruce from among a bed of roses in Berlin's Friedrichshain district, then return it once Christmas is over. The experience itself inspires a tradition of its own:

> We continued to visit the tree regularly; it has taken root again. The silver foil paper stars hung on the branches for quite some time, some even until spring. A couple of months ago I saw the tree once more. Now it's a good two stories high and as wide as medium-sized factory chimney. It seems strange to think that it was once a guest in our home.

Discoveries under the Tree

In the nineteenth century it was common to find so-called Paradise or Christ Gardens arranged under trees. This was a square wooden platform with a hole in the middle surrounded by a fenced-in "garden" of wood or moss. Some-

An unusual visit at the Christmas tree

times there was even a manger with Mary, Joseph, and their child inside. In America, where these decorations enjoyed particular popularity among the Moravians of Pennsylvania, they were called *Putz*. The small, labor-intensive theme parks under the tree came in various forms: they could be tiny lifelike landscapes created with earth and rocks carried indoors, even, at times, grottos with figures of shepherds and sheep. Some represented a farm, with a house and a barn, and maybe a winding road with hand-carved Lilliputian horses and cows treading it. There might even be little waterfalls and fountains, or metal goldfish in an artificial pond, steered by moving magnets beneath. Sometimes these miniature worlds took so much space that furniture had to be removed temporarily from the parlor.

There was also once a time when people were satisfied decorating Christmas trees with edible objects rather than less consumable gifts. It is tempting to romanticize the past, but Christmas was simpler than today, perhaps evoked stronger feelings as well. That all changed when the tree landed in the middle of a gift-giving culture that altered the face of Christmas entirely. Giving presents to friends and family is itself an old practice, with many dimensions and meanings, but the idea is not originally connected with the Christmas celebration. One could say it simply found a fitting occasion. The development of the practice of Christmas gift giving is closely tied with industrial advances that made it possible, in the second half of the nineteenth century, to produce large numbers of toys in a relatively short period of time. A children's story in *St. Nicholas* magazine from 1876 relates that Santa comes with "mysterious looking parcels," which he then proceeds to hang on the tree, one after the other. Soon the pace of Christmastime giving increased, and the presents were simply too heavy to hang on the branches. Gift hanging would have also been inconsistent with the idea of the tree decoration, which was really supposed to stay on the tree. In 1896 *Good Housekeeping* explicitly recommended that gifts not be placed on the branches of the tree; instead, one should "cover the floor immediately surrounding the tree with white paper . . . [and] arrange the gifts around the base." From then on boxes, usually wrapped in white, red, or green tissue paper, became standard—they added a different dimension, creating value in itself. Even simple, inexpensive presents, now presented in holiday dress, could be emphasized and seem to be much more.

In a certain sense the diverse presents, later with even more elaborate golden ribbons and expensive wrapping paper, came into competition with the tree, perhaps even taking away some of the significance it had borne until that time. As mass consumption burgeoned, new markets arose to exploit the Christmas celebration, and the tree itself was increasingly drawn into the growing shopping chaos: the "Christmas industry." In the process, more and more Christmas trees were erected outdoors—in private gardens as well as public squares and parks—and decorated with strings of lights. They shone through the entire advent season as a symbol of peace (and prosperity).

When the consumer society became the target of rebellious students in many European countries and in North America as well, the Christmas tree wasn't spared. The tree was a symbol of what many at the time described as the "Christmas terror," the Christian message having been largely forgotten. In Stockholm, for instance, protestors in the late 1960s targeted "tinsel hysteria" and even wanted to do away with Santa Claus.

But for Christmas as for other elements of the popular culture, everything old became new again; the holiday catered to a sensibility for which the logic of being merely "brand new" had become stale. In the 1970s and 1980s, as pale-faced porcelain dolls and all other manner of previously useless junk were being resurrected at flea markets, so did "authentic" candles and simple, handmade Christmas decorations return.

One variant of this retro trend is an emphasis on natural tree decorations. Materials such as terra-cotta, beeswax, straw, and wood are all venerated anew. The decoration of the tree goes through fashion cycles. Trees are decked out in a single color, with the corresponding colored balls and ribbons to fit the interior design. Or they are decorated according to a certain theme—here is a "moon and stars tree," there a "magic tree" with toadstools, next a "fairy tale tree" with the appropriate figurines and mythical beasts, or a "circus tree," or a "bird of paradise tree," or a "fruit tree" with apples, slices of orange, and nuts—the latter being essentially a return to the roots, a reminiscence of the time when the Christmas tree was decorated primarily with edible objects. We have seen how the American tree developed in its own particular ways and may be surprised to discover such trees in other parts of the globe. For example, driving across German landscapes around Christmas, you are likely to find sumptuously illuminated trees in the suburbs so typical for American front gardens: definitely a cultural import of a fashion to Europe, whence the original inspiration for the Christmas tree comes. Tree ornaments have been transplanted similarly.

Ideological Concerns and National Pride in the Tree

Even today in many Catholic countries it is uncommon to display a tree in the home, where the manger—traced to Francis of Assisi in the early thirteenth century—is the traditional icon of the season. Instead of private trees, these communities often raise enormous trees in public squares. A 180-foot-high "tree" is erected each year in front of the Porto city hall in Portugal—an imposing scaffold with strings of lights in the form of an evergreen. In 2009 a fifteen-ton artificial Christmas tree in Russia's Pacific port city Vladivostok made headlines after a snowstorm toppled both the tree and its expensive decorations.

Christmas trees have also been displayed in service of national interests. In England and Scandinavia it was once popular to decorate the Christmas tree with national flags and streamers. During the Franco-Prussian War of 1870–71, trees were placed in military hospitals and barracks to inspire soldiers.

During the period of Nazi Germany, the tree remained the center of Christmas activities for the families and later also for soldiers. Throughout the world, the decorated tree was the preeminent symbol of the German Christmas celebration, which lent itself readily to ideological manipulation. Thus the National Socialist propaganda machine strove to emphasize the tree's Germanic roots and to paint it as the direct descendant of the mythical tree of life—all in order to distract from the Christian meaning of the celebration. Christmas tree ornaments were renamed *Julschmuck,* or yule decoration. Angels, stars, and bells disappeared, while balls with old depictions and symbols of animals and plants were praised as "vivid archetypes of the ancient Germanic primal wisdom." Alongside these pieces, either baked or cut out on a jigsaw, were relatively plain decorations such as apples, nuts, or candles. For the top of the tree, "in place of the more or less tasteful glockenspiel of years gone by," as one observer wrote, was recommended a swastika or a Black Sun, another Nazi symbol. The historical revisionism took on such grotesque proportions that even the Christian roots of the tree were simply denied. In 1935, after an article in the *Osservatore Romano,* the semi-official organ of the Catholic Church, called into question the character of the Christmas tree as a Christian symbol and recommended that faithful Christians should build a manger instead, the German press accused the Vatican of suppressing an old Germanic tradition.

The Nazis' fascist allies in Italy ruled against the Christmas tree, but in Germany the tradition thrived. In 1937 Friedrich Rehm wrote: "Just as we cannot imagine that the entire Christmas celebration gets its deepest spiritual content in a religion which arose in the Orient, so can we not conceive that a German *Tannenbaum* could have anything to do with a manger in a stall in Bethlehem." And one didn't only sing songs around the tree; one also avowed one's faith in the Führer. Some trees even had red balls with swastikas which read "Heil Hitler."

During the Cold War, some Christmas ornaments available on the American market came under attack. These once-innocent baubles were now suspect: if they originated in such "Russian dominated lands" as Poland, Czechoslovakia, and East Germany, they contributed to the economic success of the Communist enemy, a *New York Times* editorialist wrote in November 1952. The Glassware Institute of America also decried the increase in imports. Customers were made aware "of the distinguishing characteristics of the domestic and imported products." As imports sharply decreased, the American Christmas tree decoration industry grew in size and stature.

Christmas Trees in Other Domains

When we celebrate Christmas in the Northern Hemisphere, of course, it is summer in the southern half. In those antipodal climes, the green of the tree—whether an evergreen or one that thrives only in the warm seasons—can hardly have the same associations as it does in our latitudes, even if it is covered in artificial snowflakes. Perhaps it evokes Europe or North America—the native homes of the trees.

And regardless of hemisphere, as we approach the equator, the sharp seasonal distinctions that help give the Christmas tree its iconic power disappear. A charming incongruity arises when a fir or a spruce is shipped to the tropics and then placed next to a palm tree. What does a Christmas tree mean in Southeast Asia, for example? Although most of the populations there practice Buddhism, Confucianism, or Taoism, there are also many Christians who celebrate Christmas. In Vietnam, for instance, many converted to Christianity during the French colonial period. The population of the Philippines, once a Spanish colony, is majority Christian. The Christians there also find imaginative solutions to the lack of a "Christmasy" climate. Some make "Christmas trees" from colorfully painted branches. Others modify native palm trees—or even bales of rice straw—and deck them with fanciful homemade decorations. Chinese people often use Christmas regalia as symbols of their own seasonal gift giving,

Christmas in Japan, circa 1920

without even realizing their association with Christianity. In Japan, too, the Christmas celebration—called *kurisimasu*—and the Christmas tree are used without any religious connotations, simply for their exoticism. Usually Japanese celebrants use artificial trees, painstakingly decorated with lights, origami, and small fans. Even the occasional bonsai tree might be furnished with a red ball. By the end of December all the Christmas accessories disappear. No trace of them is to be found at the Japanese New Year's festival.

Decorated evergreen trees can also be found in other non-Christian communities, freed of all of their original meaning. Visitors to Turkey, for instance, will marvel at the abundance of what may be perceived as Christmas decorations on display at the end of the year. Trees—red, yellow, and, of course, green artificial ones—are abundant, especially in the well-to-do Istanbul quarter Nişantaşı. This may be attributable in part to the area's enthusiasm for Western-style consumerism. But the tree is also connected to the family New Year's festival. "Mutlu Yıllar"—Happy New Year—accompanies the decorations in the store windows, and the tree itself is properly called *Yılbaşı ağacı,* that is, "New Year's tree." On the last day of the year Santa Claus caps outfitted with blinking lights sell like hotcakes.

Meanings are subject to change, but the Christian symbolism of the tree remains strong in the face of the obstacles it has faced in some places. In fact, that symbolism is so powerful as to be unwelcome among some populations. In Israel, for example, where the Christmas tree typically appears only among Arab Christians, it is thought to damage business for hotels and restaurants. A few years ago trees displayed by the Spanish fashion chain Zara inspired a protest campaign by Israeli Jews. As a result of the demonstrations, the trees had to give way to menorahs.

One of the most expensive Christmas trees—if not *the* most expensive—was put on display in Abu Dhabi in 2010. There, a thirty-foot-high tree was erected in a hotel lobby and decorated with pearls, gold, and gems worth $11 million. It was meant to put their Western visitors in a festive mood, but the management was charged with having done a disservice to the Christian symbol. Several guests were disturbed by the tree and said that it had nothing to do with their conception of the celebration. Of course, complaints about cultural pollution of the symbol seem ironic in a world where some Westerners trim their trees with Arabian Nights–brand balls decorated with Oriental patterns.

In an era of heightened cultural sensitivity, some yuletide celebrants are careful to refer to the "holiday tree" rather than the "Christmas tree," precisely to guard against giving offense to those who don't celebrate the Christian feast. But whatever the tree is called, the effects it unleashes, the associations it triggers, remain somewhat unpredictable. And this is also true of the sentiment that such a tree can create in various cultural contexts, if simply because its meaning is subject to changes with time and place.

Tree lights and stars in Altadena, southern California

As a Christian symbol the tree inspires a quietude, reflectiveness, and joy in places where the story of Christmas is known and prized. In contrast, where the story is not part of the culture, the tree may elicit resistance, or may simply take on layers of meaning unrelated to those that Christians attribute to it.

The World Capital of the Christmas Tree

If there is one place in the world where a public Christmas tree, erected for common delight, has gained such prominence that it might fairly be called the tree's world capital, it is New York City. Of course, the significance of a tree depends not solely on its size but also on its surroundings. In a metropolis dominated by skyscrapers, tall trees fit perfectly into the high-rise forest. At the same time, the vertical environment demands that any given tree must be large enough to be seen in the first place. In New York, a very particular symbiosis of trees and buildings prevails. The trees transform the urban environment, and the city in turn imbues these immigrants from the countryside with a certain urbanity. In many places even "normal" trees are hung with strings of lights—usually white or yellow, not red or green, to avoid confusing (perhaps slightly tipsy) motorists. Since 1945, Park Avenue has been decorated with Christmas trees, following a ceremony next to the Brick Presbyterian Church, in remembrance of the American soldiers who died in the Second World War.

Probably no city in the world offers Christmas in so many flavors as does New York. And it is no surprise that some people in this city have developed very particular tastes regarding how their Christmas trees must be decorated. One example is the American Museum of Natural History's Origami Holiday Tree, a tradition started more than thirty years ago by Alice Gray, a scientific assistant in the entomology department. In its latest incarnation, the tree displayed five hundred little folded-paper replicas of animals and objects that can be found in the museum.

Some people's tastes, of course, test the limits—in New York perhaps more than anywhere else. Outré displays in the privacy of one's home are generally safe from criticism, but public spectacles are almost guaranteed to spark fierce debate. Intense protests broke out in New York in 1998 when AIDS-prevention activists planned to decorated a tree in Central Park with condoms. Then-mayor Rudolph W. Giuliani called it "one of the most idiotic ideas" he had ever heard. Despite backing from powerful sponsors, including Coca-Cola, the plan was not carried out.

The most famous of New York's trees—and the one that inspires the loftiest expectations and the most unflagging efforts—is the one that has, since the early 1930s, graced Rockefeller Center, in the heart of Manhattan. Every year brings a quest for the most beautiful tree in the region. The manager of Rockefeller Gardens searches the area around New York in a small airplane for a suitable tree, then attempts to convince the owner that the tree's true magnificence will be achieved only when it is illuminated by tens of thousands of lights, attached to more than five miles of cable. Once the deal is sealed, even more effort is required to transport the tree to the city in a special trailer, and finally to position it with the help of a tall crane.

When to Part Ways

Whether tall enough to compete with skyscrapers or compact enough to spruce up a studio apartment, the tree has a limited shelf life. Once it begins to lose its needles, its magnificence irretrievable, the symbol of vitality succumbs to the inevitable. But exactly when the tree should go is another topic for vigorous dispute. For many the operative day is January 6, the feast of the Epiphany.

One tradition for disposal of the remains of the tree is to burn it along with other deadwood in an Easter Fire. That, no doubt, creates a satisfying spectacle, but at a time when sustainability is a concern for many, the disposal and recycling of discarded Christmas trees has taken on new importance. In this light, a Danish tradition of giving the tree second life may seem prescient: rings of suet and balls of fat are hung on the tree, which is then placed in the garden to the delight of the birds. A more common strategy in urban and suburban areas today is to use the discarded whole trees in landscaping and in gardens, where they loosen the soil and support plant growth. Dried chips from the trees can be used as fuel for wood stoves. Complete trees are used to fortify sand dunes and protect the new accumulations from rapid erosion; sunk into ponds and lakes, they form a protective zone for fish—an artificial reef—among their dead branches. Finally, the purist who has used a living, potted Christmas tree can easily transplant it outside, as long as its root system remains intact, and the planter has waited out the last hard frost.

Trees that weren't sold before the holiday can be put to special use as well. Some zoos feed them to polar bears, elephants, camels, llamas, and mountain goats. Apparently their sweet resin makes them a favorite food of elephants, and they are rich in vitamin C. However, some critics complain that trees sprayed with chemicals either before or after being felled are not suitable feed. Care must be taken to remove tinsel, too, which can contain high levels of heavy metals.

The Almost Endless Innovation of Christmas Trees

In every place where Christmas or New Year's is celebrated with trees, some once-living firs and spruces have been replaced by artificial trees, some lifelike, some abstract, some even made of inflatable plastic. Nor is this a recent phenomenon. In nineteenth-century Germany a "feather tree" was marketed: instead of branches, it bore large feathers, dyed green and meant to create the illusion of a "real" tree. Such trees were also available in the United

States, where they could be purchased with red artificial berries that functioned as candle holders. Later variants displayed blue-green, limpid-looking plastic strips that had no similarity to needles. More recent artificial trees reveal themselves to be imitations only upon close inspection, so nearly do they mimic the original, some even feeling similar to the touch. Others are unabashedly artificial, crafted of bright red, gold, blue, or white plastic. These trees do not lose their needles and are thus reusable, but most can still catch fire. To limit this risk some are outfitted with small electronic smoke detectors. And, of course, one can buy Christmas tree scent spray to complete the illusion by simulating the smell of a freshly cut tree. As a final touch, the tree can be sprayed with artificial snow.

The shape of the first fake trees was "perfect," slim and straight. But doesn't the true charm of the real thing come from its slight imperfection? When the manufacturers of artificial trees grasped this bit of psychology, another market boom followed: the more "natural," "imperfect" artificial tree. Today there are an estimated fifty million artificial trees in use in the United States, and the trend is growing: artificial trees have been outselling real ones for two decades. Some advocates claim that artificial trees are more energy efficient, but, as with every aspect of the Christmas tree, opinions differ. New studies have raised some doubt, especially since artificial trees are produced with plastic, and thus petroleum. On the other hand, manufacturers often work with recycled plastics.

Either way, most artificial trees are produced in East Asia, so they must be transported halfway around the globe to their markets. At the other end of the "natural" spectrum, buyers unwilling to have real trees in their homes that have been treated with pesticides and chemical fertilizers have created a market for "organic" Christmas trees.

The list of materials used for artificial trees is almost endless, ranging from practical to whimsical. There are foldable trees made of tin and aluminum; trees made of potatoes, chocolate, or Murano glass; trees studded with diamonds or cut crystals. Fiberglass trees in especially vivid colors, such as bright blue, are popular. A small illuminated rotating porcelain tree plays Elvis Presley's "Blue Christmas." Presley himself decorated his ranch house with a nylon tree that had red ornaments and a revolving base tootling Christmas songs. The tree in his Memphis mansion Graceland was much bigger in scale.

Inevitably, as the material of the tree became an object of experiment, some began to see the form as a medium for artistic expression. Taking the *idea* of the Christmas tree as a starting point, a self-anointed artist might create something that only slowly begins to take shape as a tree. For example, a ladder is decorated with strings of lights, or books are stacked on a shelf to mimic the conical shape of the tree, or a large number of green beer bottles are stacked into a "tree." In some churches, a "singing tree" can be viewed each year: singers with green habits are arranged in tiers on a large frame, taking the triangular

A wonderfully radiant "tree" in Toronto

form of a tree. The top is adorned with a "live angel" and a star. Is this proof of the vitality of the ever-changing tradition of the Christmas tree or simply a playful interpretation gone awry?

Occasionally a Christmas tree—or an approximation of one—is designed in a way calculated to elicit strong emotions. In 2011 a tree erected in Rome, made of white papier-mâché, caused such a stir that the city council demanded its removal the very next day. The mayor, Gianni Alemanno, said that he couldn't stand the obelisk-like thing and requested that it be replaced by a normal fir tree.

But such more or less convincing innovations are not alone in renewing the image of the Christmas tree; a seemingly contrary trend pushes us toward observing more old-fashioned customs. This yearning for the ways of times past may be typical of an age obsessed with things both retro and new. Consider, for example, the Church of Saint-Georges in the Alsace town of Sélestat. To this day they still hang the tree from the ceiling and shine but one light on it, thus lending the tree a certain simple magnificence in the darkness of the nave.

The Perfect Tree

Various kinds of trees have been most prized at different times in America as well as in Europe. After conservationists began to decry the strange landscapes left by indiscriminate clear-cutting of forests, often along the highways, some entrepreneurs discovered that the planting of Christmas trees could become a profitable business—especially on land that was not suitable for other kinds of farming. Norway spruce and balsam fir enjoyed early popularity, and during the Depression the Scotch pine, native to Europe and Asia, made its appearance. The long needles can make it challenging to adorn the trees, but their great advantage is that they do not lose their needles as quickly as most other varieties.

Today the Douglas fir (or Montana fir), with its short needles that stay on their branches in heated homes, is one of the most popular Christmas tree species. The primary challenge is to find trees that dry slowly, do not represent such a significant fire hazard, and have the desired attractive appearance. The cultivation of trees for use at Christmas has become an industry that leaves nothing to chance. There is too much money involved. Every year about thirty-five million real trees are used for this purpose in the United States, more than fifty million in Europe. In the United States, most trees are harvested from farms in Oregon, North Carolina, Michigan, Pennsylvania, Washington, New York, and Virginia. There are more than twenty thousand U.S. Christmas tree growers, and on about twelve thousand farms people can cut their own trees. Denmark, not generally associated with vast forested tracts, is the largest exporter of trees produced especially for this purpose in Europe. The small kingdom sends around ten million trees abroad each year. Planting occurs in spring. To grow a seven-foot tree from seed takes seven to ten years. A few trees are grown from the stumps of trees previously cut, but whose root systems have been left intact. Of course, Christmas tree plantations are not to be confused with genuine forests, but they still provide habitat for some wildlife—from birds to mice, squirrels, and rabbits, to deer and horses—which in turn can pose a danger to the seedlings and even grown trees. Poaching is also an issue on some plantations.

The search for the perfect Christmas tree goes on. The major challenge is to select seeds from trees that are superior in terms of shape and color. But that is not all. Plant geneticists continue to research the highly complex genotype of our evergreen trees. The difficulties they encounter in their work can be seen in one comparison. The trees have only twelve chromosomes, but each one holds around seven times as much genetic information as the forty-six chromosomes that humans possess. Evergreens are known to be especially resilient to changes in the climate, and if the genes responsible for that characteristic can be identified, cultivation can be enhanced even further. Using cloning technology, Danish biologists are trying to increase the number of trees that have a pleasing, symmetrical form, are resilient to frost, and lose fewer needles.

Other geneticists seek to directly alter the genes of the trees. Attempts have already been abandoned to increase resistance to pests via introduction of a particular gene. Some experiments sound absurd, such as an effort by British students to make trees glow by introducing luciferase, an enzyme found in fireflies. To date, their research has shed little light.

Once There Was a Christmas Tree

What we immediately recognize as a Christmas tree is the product of centuries of custom, myth, vivid imagination, craftsmanship, and iconography. Some historians have attempted to draw a direct link between heathen rituals and the Christmas tree. But while this may be a useful insight, the reality is considerably more complex. The German historian Alexander Demandt wrote this about the Christmas tree, which he characterized as the "emblem of domestic bliss":

> Its roots are many and reach far into the past: Christian, ancient, and Germanic elements, which Chateaubriand [in] 1831 saw as the composite of European culture, come together here into a rare density. The meaning is Christian, the origins are ancient, and the form of the Christmas celebration is Germanic.

Was the Christmas tree something consciously and deliberately "invented"? Was someone searching for it? Probably not, but it was found nonetheless. It would be wonderful to resurrect that moment when the tree first unfolded in its magnificence and put its viewers in a state of awe. Let us assume that the nameless inventor of the Christmas tree knew about the mythical trees of the past and the old heathen customs. According to our current state of knowledge, the Christian paradise play, with its decorated tree of life and death at the center, played a decisive role in the emergence of the Christmas tree. In addition, the use of the tree in the play might have lent particular emphasis and dynamism to the custom as we know it today. In a certain sense the evolution of the tree is but a repetition of the development of Christmas itself—a celebration that also came into the world incomplete, with roots deep in the mythologies of various Eurasian peoples, and only slowly came into its own.

Of course, the Christmas tree is a quiet triumph of man over nature. For only man has the power to change the natural habitat of the tree and to shape it according to his imagination, even if that "shaping" means only decorating it and lighting it to make it "bloom." No description could fully illuminate the imaginative combination of natural and cultural elements, could precisely decode the transformation of the tree of paradise into the Christmas tree, but perhaps this explains at least a part of the fascination the tree retains to this day. Emotion and excitement must have played important roles from the start. The splendid Christmas tree, this drama of light from the fairy-tale forest, is the magical focus of a ceremony that will continue to preserve its inner secrets in spite of all attempts to the contrary.

An anthropomorphized Christmas tree and a snowman in a rarely
seen tender moment

Everyone who has ever had a Christmas tree has carried the tradition forward and made it his or her own. Each person has played with the various elements to approach his or her idea of a perfect tree. In the process the custom has undergone astounding change and adaptation, and everywhere in the world where the tradition is alive today, further evolution continues. Mysterious and ancient though its roots may be, the Christmas tree remains one of our more visible icons, and it is always being invented anew.

Selected Bibliography

Albers, Henry H., and Ann Kirk Davis. *The Wonderful World of Christmas Trees*. Parkersburg, Iowa: Mid-Prairie, 1997.

Andersen, Hans Christian. *The Fir Tree*. Available in many editions.

Breuer, Judith, and Rita Breuer. *Von wegen Heilige Nacht! Das Weihnachtsfest in der politischen Propaganda* (Holy night? No way! Christmas in political propaganda). Mülheim an der Ruhr: Verlag an der Ruhr, 2000.

Cullmann, Oscar. *Die Entstehung des Weihnachtsfestes und die Herkunft des Weihnachtsbaumes* (The emergence of the Christmas festival and the origin of the Christmas tree). Stuttgart: Quell-Verlag, 1990.

Demandt, Alexander. *Über allen Wipfeln. Der Baum in der Kulturgeschichte* (Over every treetop: The tree in cultural history). Cologne: Böhlau, 2002.

Elm, Hugo. *Das goldene Weihnachtsbuch* (The golden Christmas book). Halle: G. Schwetschke, 1878.

Herrlein, Theo. *Das Weihnachtslexikon. Von Aachener Printen bis Zwölfernächte* (The Christmas lexicon: From Aachener Printen to the twelve nights). Reinbek bei Hamburg: Rowohlt, 2005.

Klimke, Karl. *Das volkstümliche Paradiesspiel und seine mittelalterlichen Grundlagen* (The traditional paradise play and its medieval foundations). Breslau: M. and H. Marcus, 1902.

Kronfeld, Ernst Moritz. *Der Weihnachtsbaum. Botanik und Geschichte des Weihnachtsgrüns, seine Beziehung zu Volksglauben, Mythos, Kulturgeschichte, Sage, Sitte und Dichtung* (The Christmas tree: Botany and history of the Christmas tree, its relation to popular belief, myth, cultural history, legend, custom, and poetry). Oldenburg, 1906.

Lauffer, Otto. *Der Weihnachtsbaum in Glauben und Brauch* (The Christmas tree in belief and tradition). Berlin: Walter de Gruyter, 1934.

Mann, Thomas. *Buddenbrooks: The Decline of a Family*. Trans. H. T. Lowe-Porter. London: Vintage, 1996.

Mantel, Kurt. *Geschichte des Weihnachtsbaumes und ähnlicher weihnachtlicher Formen. Eine kultur- und waldgeschichtliche Untersuchung* (History of the Christmas tree and similar Christmas forms: An investigation of culture and forest history). Hannover: Schaper, 1975.

Marling, Karal Ann. *Merry Christmas! Celebrating America's Greatest Holiday*. Cambridge: Harvard University Press, 2000.

Moser, Dietz-Rüdiger. *Bräuche und Feste im christlichen Jahreslauf. Brauchformen der Gegenwart in kulturgeschichtlichen Zusammenhängen* (Customs and festivals in the Christian calendar: Customs of the present in the context of cultural history). Graz: Edition Kleidoskop im Verlag Styria, 1993.

Otto, Sigrid. *Oh, Tannenbaum. Zur Geschichte des Weihnachtsbaums* (Oh, Tannenbaum: On the history of the Christmas tree). Berlin: Staatliche Museen Preußischer Kulturbesitz, 1992.

Rapsilber, Maximilian. *Weihnachtszauber. Ein deutsches Haus- und Familienbuch* (Christmas delight: A German house and family book). Berlin: Simon, 1912.

Rehm, Friedrich. *Weihnachten im deutschen Brauchtum* (Christmas in German customs). Leipzig: Strauch, 1937.

Restad, Penne L. *Christmas in America: A History*. New York: Oxford University Press, 1995.

Schnurre, Wolfdietrich. *Die Leihgabe* (The loan). Berlin: Aufbau, 2010.

Stokker, Kathleen. *Keeping Christmas: Yuletide Traditions in Norway and the New Land*. Saint Paul: Minnesota Historical Society, 2000.

Waits, William B. *The Modern Christmas in America: A Cultural History of Gift Giving*. New York: New York University Press, 1993.

Illustration Credits

Ingeborg Weber-Kellermann, *Das Weihnachtsfest. Eine Kultur- und Sozialgeschichte der Weihnachtszeit* (The Christmas celebration: A cultural and social history of Christmas time) (Lucerne: Bucher, 1978)   viii, 6, 20, 32, 94

The Árni Magnússon Institute for Icelandic Studies   9

Hugo Elm, *Das goldene Weihnachtsbuch* (The golden Christmas book) (Halle: Schwetschke, 1878)   24

Hans Christian Andersen, *Der Tannenbaum* (Cologne: Middelhauve, 1984)   47

Winterthur Library   53

John F. Kennedy Presidential Library and Museum, Boston   59

Maximilian Rapsilber, *Weihnachtszauber. Ein deutsches Haus- und Familienbuch* (Christmas delight: A German house and family book) (Berlin: Simon, 1912)   68

Emil Schiller, *Weihnachten in Japan und China* (Görlitz: Hoffmann and Reiber, 1920)   76

All other illustrations are vintage postcards or sheets or photographs by the author.